HIDING IN PLAIN SIGHT...

Hiding In Plain Sight...

There is "Something" up there, and it is like nothing you have ever seen... Until now...

SCOTT GEAREN

Scott Gearen LLC

CONTENTS

To my second grade teacher, Ms. Harrell, who took me to a planetarium when I was seven years old...

I thought it was a planetarium...

First Printing, 2020

Foreword

This is not an ordinary book, it is not a great work of literature, it is not a novel, a biography, or science fiction... this is *science-nonfiction*.

The pictures displayed in this book have never before been published anywhere or in any form of media.

My hypothesis, and comments along with the pictures presented in this book may cause you to contemplate everything you think you know...

Either way, you will not look at the sky above you the same way as you did before... This photographic "*evidence*", never before seen, could be a life changing experience for you.

You have been warned...

This is the first and only book to document, with 100 percent authentic, verified and validated, photographic evidence of an...

Unidentified Flying Object

| 1 |

Not Everything is a Weather Balloon

Warning!

What you will see in the next few pages may forever change the way you look at our Universe.

Anything you see in the air that you cannot identify can be labeled an Unidentified Flying Object, a UFO. Because you could not identify the object does not automatically imply it is an alien spacecraft from another planet located in a distant galaxy on the other side of our universe... but it might be... if you only had a camera...

I did see a UFO, and I had a camera. I was able to photograph and capture an incredible series of pictures never seen before. All pictures presented in this book were taken by me on 30 April 2019, they are property of Scott Gearen LLC, are copyright protected, and may not be reproduced without prior written consent.

These pictures are presented as evidence of what I witnessed and captured with my camera. I saw it, I took pictures of it, and I do not know what it is. I have not seen any pictures anywhere that are as clear or defined as these, but even with the best pictures in the world, this object is more of a UFO than ever. Even though the pictures of this UFO have never been made public, I believe someone, somewhere has seen it before and may know what it is.

Since I was able to capture these pictures on my personal 35 mm camera, I believe the United States government, with the most advanced technology available, is aware of this UFO, and it is possible there are government employees or contractors operating it, either alone or in conjunction with Extraterrestrial Beings (ETB).

The pictures included in this book and others like them, when reported to authorities are most often referred to as "weather-balloons" ... among other things. I have seen weather balloons and I have never seen a weather balloon that resembles any of these pictures. If there are weather balloons like this... where are they and why isn't one like this produced by the people that claim it is a weather balloon?

These pictures have never before been published; anywhere. There may be other pictures of this UFO kept locked in secret files by the governments of the world, but this is the first documented proof that *something* other than birds and planes are in our atmosphere.

Something very strange indeed...

| 2 |

View from the 22nd Floor

Always keep a camera close enough to use when you see something strange, especially if it you can't identify it

I was in Panama City Beach, Florida because I had been asked to speak to a group of High School Seniors scheduled to graduate. Prior to going to the scheduled event, I was doing push-ups on the floor in my hotel room. I was on the 22nd floor facing East, with an unobstructed view of the sky. It was not unusual for me to do push-ups in hotel rooms, especially this year as I had a set a goal for myself to do 100,000 push-ups over the course of the year, I had 10 sets of 50 push-ups to complete to stay on schedule.

It was while I was doing push-ups on the hotel balcony that I noticed something shinny in the sky that seemed inordinately bright. I first thought it was a reflection from an airplane passing overhead, but I quickly realized the bright object was not moving as an airplane would and appeared to be stationary. My next thought was it must be a helicopter hovering, I continued to observe it and after just a few seconds I realized it was not a helicopter, that it was something very unusual, and I began taking pictures and videos with both my iPhone, and my 35mm camera which has a 55-210 mm zoom lens.

I knew the object was unusual, but the thought that it was a UFO did not enter my mind. I wish it had, I would have focused on it and taken more pictures and video. While I was observing the UFO and taking pictures I wondered if anyone else was aware of this UFO and I looked at the crowds on the beach to see if anyone was looking up, everyone that I could see below me was busy with beach activities and I did not notice anyone looking up towards the UFO.

I did not check the pictures on the camera right away, I just continued to take pictures until the UFO finally disappeared. It was not until about three weeks later when I downloaded them to my computer and realized I had captured something extraordinary, something very extraordinary...

I did not know what to do with these pictures, I knew they were nothing like any of the UFO's presented on television shows, portrayed in movies, or on the internet. I contacted several prominent news organizations, and well known personalities, some replied without seeing the pictures and said not interested, and others did not reply at all.

The one person that did reply was Mr. Nick Pope. I reached out to him because he often speaks about extraterrestrial events, such as the UFO incident in Rendlesham Forest in December 1980. I am a retired USAF Pararescueman (PJ) and I was stationed at RAF Woodbridge/ Bentwaters with the 67th Aerospace Rescue and Recovery Squadron (ARRS) in 1980. To stay fit for my job I would often jump the fence and run miles through Rendlesham Forest rather than on the roads.

I shared several pictures with Nick, he said "I hope that your posting this material smokes out some informed views that will help get to the bottom of all this."

Eventually I submitted a report of my sighting to two UFO organizations, I submitted a single picture with the report, which was picked up by many organizations, some shared my report along with the picture (#996) on their own websites. Several organizations located the video I took of this UFO on YouTube and added the video to their websites, such as British Earth and Aerial Mysteries Society, who present an interesting hypothesis on their website, and speculate this UFO is a "Crypto-Creature".

I currently do not know if the report I submitted to these organizations will ever be investigated, there is criteria in place by the organizations to conduct further investigations, and clearly my pictures meet the established criteria. I was told the pictures I submitted were looked at by several analysts; one analyst supposedly said "I almost had a heart attack" when he saw them. I can relate that to my own feelings when I first saw them.

Additional comments by an "expert" in this field, and who does not wish to be identified are:

Anonymous: "I think you saw an advanced proto-type aircraft with stealth and cloaking abilities". "I'm willing to bet NONE of the tech was invented by Humans. I'd go as far as saying it's either a recovered crash vehicle we finally figured out how to operate...or it's development was with the blessing and assistance of ET and his friends."

Anonymous: "My personal opinion is you have captured something very unique. Something that straddles 2 worlds. I do believe the craft is OURS... but I also believe it is of future tech...or "foreign technology" ... which is a HUGE field according to the USAF/USN. It CAN include Alien technology under "foreign technology".

There was, and still is, much speculation about what the object is... it has not yet been officially identified. It may never be officially identified, but for now it is officially *unidentified.*

Some won't believe these pictures are real, others will think they are faked somehow, and those that do know what it is will most likely identify it as a "weather balloon; or remain silent and not say anything, and hope this goes away.

One of the things I learned from these pictures is UFO's are not all going to be the stereotyped image of the "flying saucer" which has been ingrained in our minds from the first reported sighting many years ago; they can appear anywhere and anytime... even in the middle of the day with a clear blue sky, and you will never see them...

These pictures are evidence UFO's are real, they are here, and unless they want to be seen, we will not see them due to their ability to change shape, and camouflage themselves with the surroundings.

| 3 |

My Pictures Don't Lie

Look at the pictures and come to your own conclusions...

One day the truth will reveal itself...

There were times while I was taking these pictures the UFO became invisible without the aid of the polarized sunglasses I was wearing, although I was not sure if my camera was able to capture the image of the UFO, I continued to take pictures. Those pictures without the image of the UFO are not included; the UFO is there, it is just not visible to ordinary eyesight or picked up with a plain lens on my camera, and those pictures would appear to be an empty blue sky, although the UFO was not always visible to human eyesight, it might be visible to animals, birds, fish or Extraterrestrial technology.

To maintain the continuity of the entire series the pictures are presented in the order in which they were taken. The respective number for each picture is included, along with the time intervals between the pictures, pictures not shown are identified by number only. This information is derived from the metadata automatically saved in the image at the time each picture was taken.

The order of the series is important because the UFO is constantly in motion. The 35mm pictures begin with picture 996 and continue in the order they were taken. The UFO stays in the same general area, but is constantly changing its appearance and shape, it disappears, it camouflages against the blue sky, it spins at an incredible high rate of speed, it glows intensely, it splits into two objects, and eventually departs in an entirely different shape than when I first noticed it and began taking pictures.

The UFO appears to be operating under intelligent control; either a human being, an extraterrestrial being, independently as artificial intelligence or it is a cryptologic creature undergoing a metamorphosis just as a caterpillar would to become a butterfly... It is a true Unidentified Aerial Phenomena (UAP).

Regardless of what it is, it is not a "weather balloon" or a space blanket floating in the air... All other options are on the table...

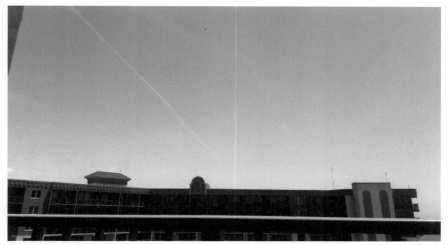

Scott Gearen

This picture was taken with my iPhone, without the zoom feature being used, to show the relation of my position on the balcony to the UFO. I do not believe the contrail in the picture has anything to do with the UFO.

The UFO is in this picture, see if you can locate it before turning the page.

Scott Gearen

In case you were not able to locate it...

It is the small bright object at Top-Center of the picture.

I estimated the UFO to be approximately 15 - 20 miles away and about 15,000 to 20,000 feet above the ground.

The following pictures have not been altered in any manner; each picture has only been enlarged so the UFO can be viewed by the reader.

The remainder of the pictures were all taken from the same spot on the balcony using my 35 mm camera with a 55 – 210 mm lens, each photo was shot at 210 mm.

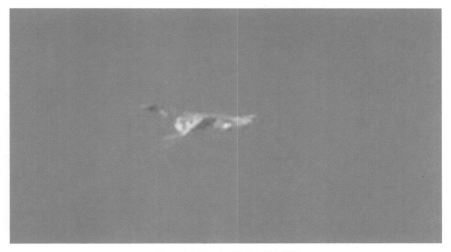

Picture 996 - 11:42:56 AM CST
Scott Gearen

This picture appears to be the most familiar due to the "airplane" shape, however, it is not an airplane. This is the one and only picture I submitted with my UFO report, it was picked up by multiple organizations and has been circulated online with much speculation as to what it is... or is not.

Notice the dark round spot on the lower right side of the object, to me, it looks like an "eye" looking down, and directly across from it on the opposite side there appears to be a "bulge", could that "bulge" be another round dark area exactly as it appears on this side of the UFO? Could this be two eyes?

The rounded area between the two "eyes" resembles the head of a Stingray with an eye on each side of its head. More commonly this is associated with a similar species, the Hammerhead Shark, but this one has "wings"...

Picture 997 (not shown) is unrecognizable.

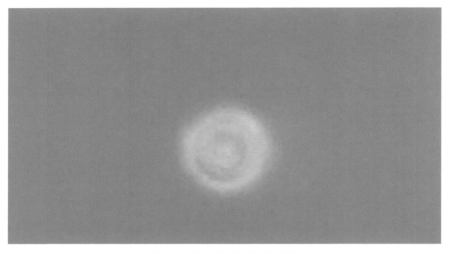

Picture 998 - 11:43:08 AM CST
Scott Gearen

Picture 998 taken 12 seconds after 996, this picture clearly shows the UFO has created this phenomenon.

There are similar pictures like this one that others have captured on their cameras and shared online. The explanations generally given are that it is an *anomaly* of the camera. This is not an anomaly. This is not a one and done picture, this phenomenon occurred several times during the series of pictures I captured.

Could this be a portal to another dimension? The thought crossed my mind, the UFO is no longer there, and only what resembles an opening...

"Black Holes" may not be "Black" after all, and this UFO is going in and out of one Dimension into another, or Galaxy to Galaxy or to the other side of the Universe...

My observation of this occurrence, and photographic evidence, validates this is part of the actions of the UFO, and previous pictures by others only captured one part of the sequence.

Picture 1001 - 11:43:22 AM CST
Scott Gearen

Picture 1001 taken 12 seconds after picture 998 appears to be blurry, however, I believe the blurriness of the UFO is not due to camera focus, I believe this picture, and others that are blurry are due to the actions of the UFO itself.

The UFO is not moving horizontally across the sky, the blurriness is due to the movement of the UFO.

It may be emitting some type of energy, or "gas" that distorts the air around it, or it might be in some type of transformation from camouflaging, "cloaking" itself from being visible to the light spectrum's we are able to see, or it might be all of them combined.

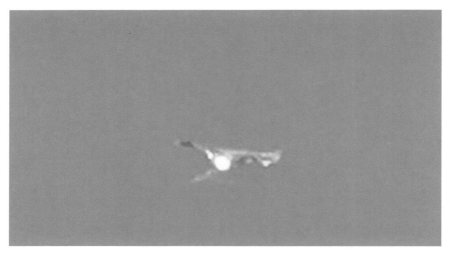

Picture 1002 - 11:43:26 AM CST
Scott Gearen

You can clearly see that in only 4 seconds from the previous picture the UFO is visible and has gone back to the shape seen in picture 996.

Another area of significance is the bright spots, they are clearly not a reflection from the sun, instead they appear to be a power source that is fluctuating. This movement of the bright areas is clear in the video that has been uploaded to YouTube.

A modern airplane would have to travel at a high rate of speed traveling horizontally to stay aloft and not come crashing to the ground.

How is this UFO defying gravity and staying in the air with no visible forward movement?

Pictures 1003/4/5 are unrecognizable and are not included.

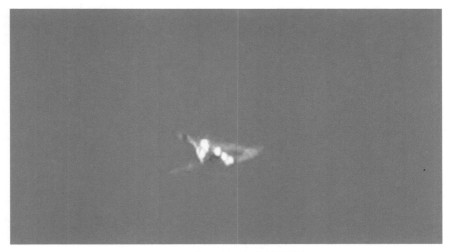

Picture 1006 - 11:44:18 AM CST
Scott Gearen

Picture 1006 is taken almost one minute after 1002.

The bright areas have gotten smaller in size and increased in numbers. These bright areas continue to move to different areas of the UFO in what appears to be an unorganized pattern.

The bright areas are not changing positions or intensity for no reason, I just don't understand the reason, or have any idea what it is.

The UFO continues to retain the same general shape.

Picture 1007 - 11:44:24 AM CST
Scott Gearen

Picture 1007 taken only six seconds after picture 1006.

Clearly the UFO has transformed itself into what appears to be one extremely bright spherical UFO with what appears to be a translucent spherical object next to it.

Is this UFO separating into two or more UFO's, traveling between dimensions, or is this part of its metamorphosis?

The bright spots on the UFO in the preceding picture, and others, seem to go through an increase in size, numbers, and intensity, as if generating heat, and ultimately the UFO is completely absorbed into a single blindingly bright, spinning, spherical UFO, and then suddenly it transforms itself into another slightly different shaped UFO.

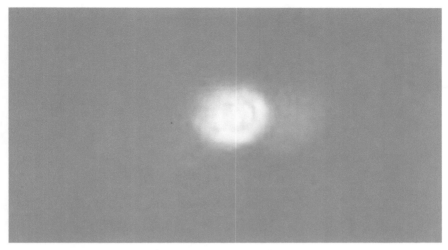

Picture 1008 - 11:44:32 AM CST
Scott Gearen

Picture 1008 is taken eight seconds after picture 1007.

The phenomenon is becoming more circular and "rings" are becoming apparent. There is still a distinctive translucent shadow to the right side of the UFO.

The UFO is clearly undergoing a highly dynamic maneuver. Why it is doing this is not easily understood with our "advanced" technology and understanding of everything we currently know, or think we know...

Our advanced technology is beginning to look like primitive technology compared to this UFO.

Picture 1009 - 11:44:38 AM CST
Scott Gearen

Picture 1009 taken six seconds after picture 1008.

The UFO is no longer an indistinguishable circular bright UFO, it is now transformed back to the same general shape, the bright sections continue to move in random patterns on the UFO with increasing levels of intensity and a myriad of shapes.

Once again, the round dark "eye" on the lower right and the "bulge" on the opposite side are apparent... and so is a "wing"...

In approximately 30 seconds the UFO did something unbelievably amazing...

and it is only getting started...

Pictures 1010 and 1011, not shown, were taken 10 and 14 seconds after picture 1009.

I could see the UFO with my just my regular eyesight, and at times only while wearing my polarized sunglasses.

I did not have a polarized lens for my camera and does not appear to have captured the image of the UFO when this occurred.

Further analysis may reveal the UFO is in these pictures.

Picture 1012 - 11:45:12 AM CST
Scott Gearen

Picture 1012 was taken 12 seconds after picture 1011.

The bright areas have changed shapes, are in different positions on the UFO, and the intensity continues to vary.

Note the bright areas as well as some areas of the UFO have straight line edges.

One of the first things taught in survival training is... "Nature does not make straight lines".

Do these straight lines imply this is not "nature" and it would therefore be "manufactured" by someone or something?

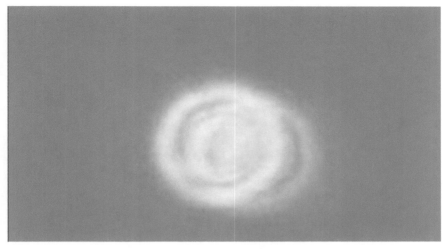

Picture 1013 - 11:45:46 AM CST
Scott Gearen

Picture 1013 was taken 46 seconds after picture 1012.

This phenomenon has clearly defined and distinct concentric circles. This may be the UFO spinning at an incredibly high speed, or could this be a portal to another dimension, and this UFO is traveling between dimensions?

Or, like a jet leaves a contrail as it moves across the sky, this UFO has used an energy source to propel itself at a high rate of speed and has left these contrail "rings" from where it departed.

Could this be "Warp Speed"?

Picture 1014 - 11:45:52 AM CST
Scott Gearen

Picture 1014 was taken six seconds after picture 1013.

The UFO is becoming blurry, and the bright areas appear to be splitting into two sections. In subsequent pictures it does appear the UFO is splitting itself into two separate UFO's.

The round dark "eye" is now bright and circular in appearance.

Why?

What is this thing?

What is it doing?

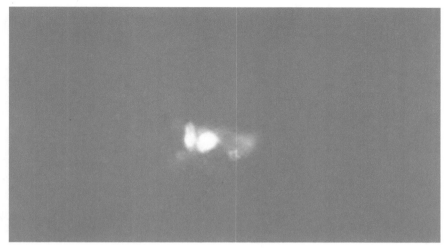

Picture 1015 - 11:46:00 AM CST
Scott Gearen

Picture 1015 was taken eight seconds after picture 1014.

The bright areas have changed their shape slightly and are more distinct, although the UFO still appears to be one UFO the two bright areas are distinctly separate and appear to be moving away from each other.

The UFO is undergoing a metamorphosis, the changes are happening at a high rate of speed and barley captured by the camera.

After viewing all the pictures, the denouement will remind you of the transformation that takes place when a caterpillar becomes a butterfly...

Could this be what is happening?

Picture 1016 - 11:46:22 AM CST
Scott Gearen

Picture 1016 was taken just 22 seconds after picture 1015.

The UFO was mostly a blur just 22 seconds prior as it was undergoing some type of shape shifting transformation.

I do not know what I am witnessing while taking these Pictures, however, I do know it is something incredible that has never been documented and presented like this.

Is this a Metamorphosis?

It looks like it based on before and after pictures. I do not know, but I did capture it on my camera. These pictures do not lie.

It appears the UFO is morphing into something...

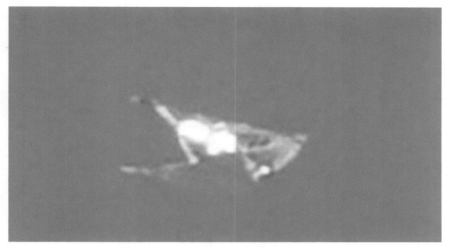

Picture 1017 - 11:46:24 AM CST
Scott Gearen

Picture 1017 was taken two seconds after picture 1016.

There has been minimal change in the shape of the UFO, but in two seconds you can see the bright areas have shifted both their shape and location, and several dark spots appear.

It appears this UFO has two "wings"; like the wings of a butterfly or a stingray found in oceans around the world. From the angle of the photograph the underside of one wing is exposed, it is as blue as the sky around it, while the top side of the opposite wing, where the bright areas are, is clearly seen. I am guessing similar bright areas are occurring on the tops of both wings.

The shape of the UFO makes it aerodynamic as well as hydrodynamic...

This UFO can move at ease in the atmosphere and potentially in the oceans as well.

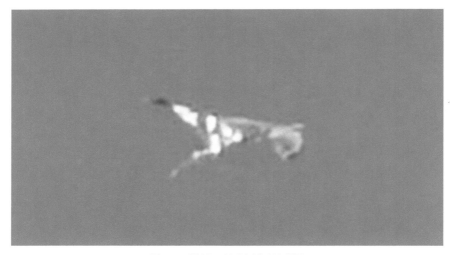

Picture 1018 - 11:46:32 AM CST
Scott Gearen

Picture 1018 was taken eight seconds after picture 1017.

The shape is still similar, and the bright areas continue to morph into different shapes and locations on the UFO. The underside is as blue as the sky around it. Looking up from the ground all you would see is the blue underside and never know it was there.

Incredibly this is the same techniques many creatures in the ocean, such as the Cuttlefish, are capable of, and in an instant, it can blend in with it's surroundings and appear to disappear, camouflaging itself so it is undetectable.

As well as the bright areas, there are other areas on the UFO changing colors, in this picture you can see changes occurring along the edges. Slight hues of blue that appear to be almost fluorescent.

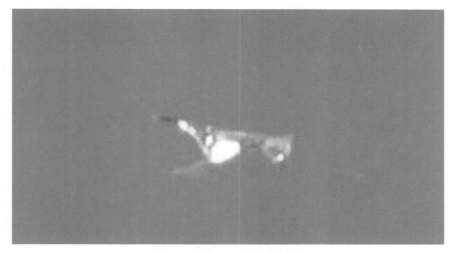

Picture 1019 - 11:46:42 AM CST
Scott Gearen

Picture 1019 was taken 10 seconds after picture 1018.

It seems the primary changes from the previous picture are the bright areas on the top side of the UFO.

The bright areas appear to be in continuous motion, increasing in intensity, moving in a random and irregular pattern, and becoming larger and larger until completely consuming the entire UFO into a singular spherical ball of intense light, followed quickly with a return to an object that is similar to where it started.

Then beginning the cycle again, and again, and again...

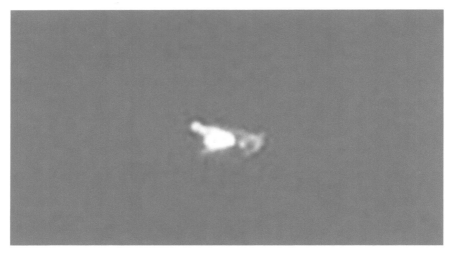

Picture 1020 - 11:46:52 AM CST
Scott Gearen

Picture 1020 was taken 10 seconds after picture 1019.

The UFO is beginning another round of it's metamorphosis.

It appears the shape of the UFO is being engulfed by the increased size and intensity of the bright area.

The blurriness of these pictures is not due to the camera.

The UFO is undergoing a procedure that is occurring so fast the camera is incapable of freezing it in a single picture.

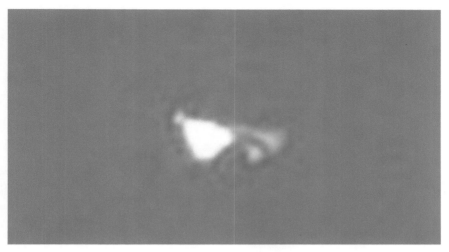

Picture 1021 - 11:46:56 AM CST
Scott Gearen

Picture 1021 was taken four seconds after picture 1020.

The bright areas seem to increase in intensity and size, becoming larger and larger until they are joined together and the UFO is completely engulfed.

Such as a caterpillar in a cocoon, while it undergoes its metamorphosis into a different creature.

A circular shape, a spiral appearance, is beginning to form in the center area of the UFO, this may be the prelude to the circular spirals seen in several pictures in this series.

Is this an anomaly of the camera? Emphatically the answer is NO.

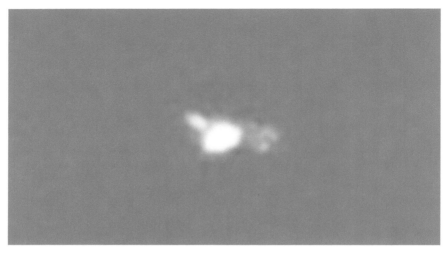

Picture 1022 - 11:47:06 AM CST
Scott Gearen

Picture 1022 was taken 10 seconds after picture 1021.

The bright areas have not completely engulfed the UFO, but, with this UFO a lot can happen in just a few seconds, which was about the time it took for me to take a picture and then locate the UFO for another picture.

The camera is in focus, it is the UFO that is creating movement within itself at an incredibly fast speed, too fast for the shutter to capture it clearly.

Ironically, even with such high speed the UFO is not traveling horizontally or vertically, it is not moving across the sky, it appears to be almost stationary.

Amazingly it stayed in the same general area for several hours, I do not know how long it was there before I noticed it.

Picture 1023 - 11:47:18 AM CST
Scott Gearen

Picture 1023 was taken 12 seconds after 1022.

In approximately 36 seconds, from picture 1019 to 1023, the UFO did something unbelievably amazing and appears to be undergoing a metamorphosis.

The cycle with the bright areas increasing in size and intensity, ultimately taking over the entire UFO, and then reappearing in a slightly different shape each time continues...

This pattern will repeat itself several times over the course of this series of pictures.

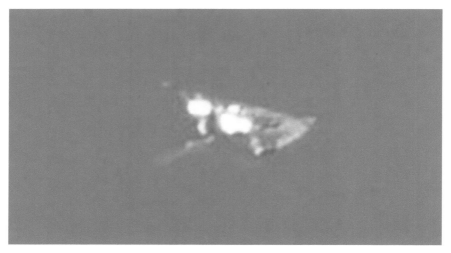

Picture 1024 - 11:47:30 AM CST
Scott Gearen

Picture 1024 was taken 12 seconds after picture 1023.

There is not any discernible change in the shape, there are still bright areas of varying intensity on the upper side of the UFO, while the bottom remains the color of the blue sky around it.

The one area that draws my attention is the bright area in the upper left of the UFO.

The bright area appears to be the shape of a perfect rectangle. Does this "rectangle" indicate something that was manufactured?

If it was, who or what did it?

Four straight lines... As mentioned earlier... here on earth...

Nature does not create straight lines...

Picture 1025 - 11:47:42 AM CST
Scott Gearen

Picture 1025 was taken 12 seconds after picture 1024.

The cycle of the bright areas increasing in size and intensity has been consistent throughout the entire series of pictures.

Once again it appears to be starting another round of activity of an increase in size and intensity of the brightness.

The area I perceive to be the leading edge, the area where the black "eye" is located, and around the upper edges appear to have a luminescent glowing color, as well as an oblong dark color adjacent to the bright area...

This UFO appears to have the same type of characteristics as a Rainbow Jellyfish, creating its own glowing neon colors as it floats through the atmosphere, as a jellyfish would in the ocean.

Is this a Bio-luminescent Creature never before seen?

Picture 1026 - 11:47:52 AM CST
Scott Gearen

Picture 1026 was taken 10 seconds after picture 1025.

The UFO continues to appear to move at a high rate of speed but is not actually traveling, it appears to remain in the same general area, the same way a helicopter would remain stationary if it was hoovering.

This is clearly not a helicopter. The bright areas seem to increase and decrease in size and intensity as if they are a power source for this UFO.

Regardless of it being a source of power or another chemical reaction, it is apparent this is part of a cycle of its continued metamorphosis.

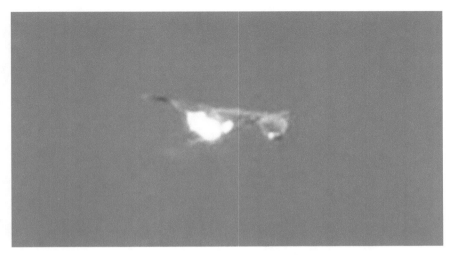

Picture 1027 - 11:48:02 AM CST
Scott Gearen

Picture 1027 was taken four seconds after picture 1026.

The bright areas are intensifying and beginning to join into one area on the UFO, once again the cycle seems to be the bright areas intensify and engulf the UFO.

After each cycle, the UFO seems to have undergone minimal changes with no clear reason what this is or why it is exhibiting a behavior that is clearly alien to anything on Earth.

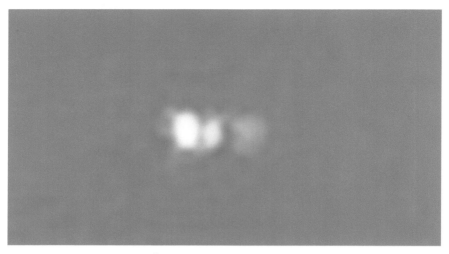

Picture 1028 - 11:48:10 AM CST
Scott Gearen

Picture 1028 was taken eight seconds after picture 1027.

The UFO is fully engulfed by the bright areas; again, it appears the UFO is splitting into two separate UFO's.

The video I took at the same time as these pictures shows the UFO does split into two UFO's.

Regardless of why it is doing this, the cycles seem to be faster and closer together as this UFO undergoes a series of changes...

Is this UFO Extraterrestrial Technology or is it Biologic...

Possibly a previously unknown "CryptoCreature"

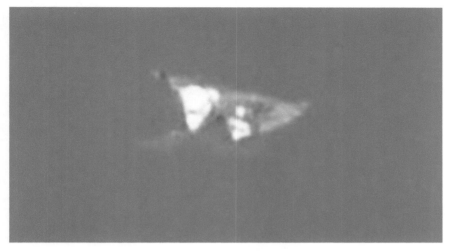

Picture 1029 - 11:49:48 AM CST
Scott Gearen

Picture 1029 was taken one minute and 38 seconds after picture 1028.

Based on the previous cycles lasting approximately one minute for the UFO to transition from the winged shaped UFO to a single spinning ball of energy, the UFO may have undergone one or more cycles since the last picture.

Two areas stand out to me in this picture, one is the bright area on the upper left of the UFO, there seems to be two circles, one large and one smaller that appear to be brighter than the area around it.

Another odd area is the small bright circle located in the middle area of the UFO... Where it looks like a head on someone's shoulders.

What in the world... no, what in the *Universe* is this UFO?

Picture 1030 - 11:49:58 AM CST
Scott Gearen

Picture 1030 was taken ten seconds after picture 1029.

The UFO is completely engulfed in a spinning, spherical shape that resembles a small sun.

It appears the right side of the circular object is separating from the main body of the UFO and a faint outline of something can be seen adjacent to it.

With our limited knowledge of the world we live in and the universe that surrounds us, there is no way we are able to fully comprehend and understand what this UFO is doing.

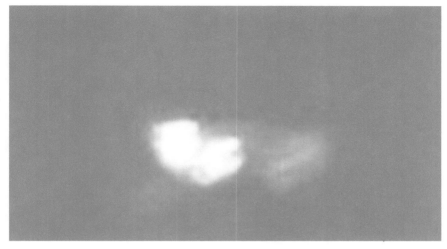

Picture 1031 - 11:50:18 AM CST
Scott Gearen

Picture 1031 was taken 20 seconds after picture 1030.

It is apparent, based on this picture, there are two distinct bright circular areas, and it shows clearly the faint and disappearing outline of something that appears attached to the right side of the two bright areas.

The UFO is somehow making itself invisible as it goes through these transitions. This translucent stage takes on the appearance of a large jellyfish, which ironically has been described as looking like an "iridescent spaceship".

Additional Pictures, and video clearly show this object is "cloaking" itself, intentional or not, this UFO is able to change shape, move at will, and manipulate itself so the human eye, and a non-filtered camera lens is unable to detect it.

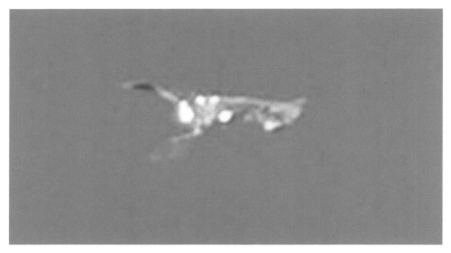

Picture 1032 - 11:50:28 AM CST
Scott Gearen

Picture 1032 was taken ten seconds after picture 1031.

The bright areas have drastically decreased their size and intensity, and the same general shape of the UFO is once again present.

Although not as evident in the pictures displayed here, the colors are continually changing, they are clear and vibrant and can be viewed in greater detail when analyzed on a large electronic screen.

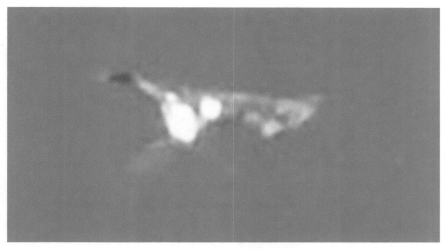

Picture 1033 - 11:50:36 AM CST
Scott Gearen

Picture 1033 was taken eight seconds after picture 1032.

The bright areas continue to increase in size and intensity and are in nearly the same position on the UFO as they were in the previous picture.

The colors have changed slightly and appear to be blending into each other, is this a camouflaging technique or is it a chemical reaction as the solid turns to a gaseous state, or something else...

Picture 1034 (not shown) was taken at 11:50:44 AM CST

Picture 1034 was taken eight seconds after picture 1033.

I could see the UFO, but the camera lens did not pick it up. This would happen several more times while I observed the UFO.

It became apparent to me as the UFO went through a transitional phase that there were times when I was only able to view it while wearing sunglasses with polarized lens.

Without the same polarization on the lens it appears in the normal picture that the UFO was not captured. It may be possible with advanced media technology that further analysis will reveal what my eyes could not see.

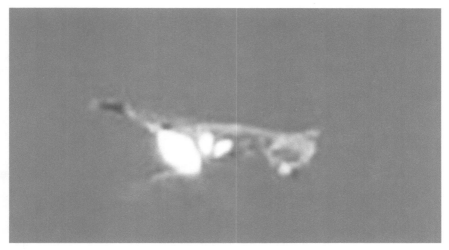

Picture 1035 - 11:50:56 AM CST
Scott Gearen

Picture 1035 was taken 12 minutes after picture 1034.

The large bright area has presented itself like this previously and generally occurred just seconds before the UFO is engulfed into a spherical, extremely bright object that appears to be spinning at an incredible high rate of speed.

The period between these singular bright balls of energy seems to be a "resting" phase for the UFO as it is not displaying the blurriness which allows me an opportunity to capture a clear image on my camera.

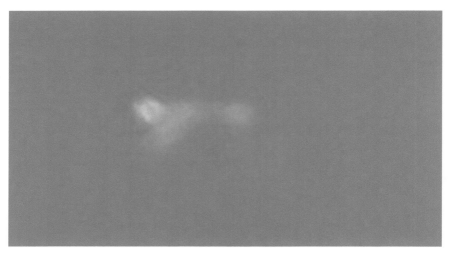

Picture 1036 - 11:51:18 AM CST
Scott Gearen

Picture 1036 was taken 22 seconds after picture 1035.

The entire UFO has practically disappeared, with only a faint outline of the outer edges and a single bright glowing area remains on the upper left of the UFO.

At first glance it is easy to think the camera is out of focus, but in actuality the UFO is disappearing as it goes through a metamorphosis.

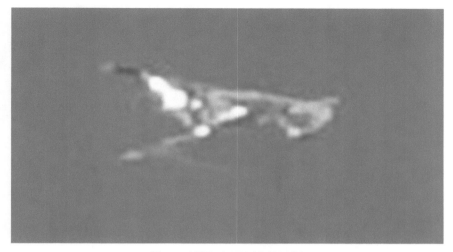

Picture 1037 - 11:51:26 AM CST
Scott Gearen

Picture 1037 was taken eight seconds after picture 1036.

Incredibly the entire UFO is visible again. The bright areas have reduced their intensity and are now in different locations on the UFO.

The body of the UFO is changing colors as well, there are shades of blue, and purple which seem to be pulsating over the UFO intermixed with the bright areas, almost like inert gases or fluids moving through the UFO.

There is once again a similar square "box" shape in the upper left area of the UFO; with four straight lines. Have you ever seen nature produce anything that resembled a square "box"?

Picture 1038 - 11:51:36 AM CST
Scott Gearen

Picture 1038 was taken ten seconds after picture 1037.

The outline of the UFO is clear and defined, it appears to be composed of both straight and contoured lines, however it is difficult to know it's composition, how it is put together, or...
how is it able to defy gravity.

One area that stands out is the blue color to the underside of the UFO.

There appears to be a section missing, but the outline of the object is visible and extends below the main body of the UFO.
It's not missing... it's just not visible.

Is this a reflection of the sky, is the UFO able to cloak itself, or is this another example of it's ability to camouflage itself enabling it to blend in with it's environment to avoid detection?

The following eight pictures: 1039 through 1046 (not shown) were taken over approximately five minutes.

Picture 1039 taken at 11:52:18 AM CST
Picture 1040 taken at 11:52:56 AM CST
Picture 1041 taken at 11:53:56 AM CST
Picture 1042 taken at 11:55:00 AM CST
Picture 1043 taken at 11:55:12 AM CST
Picture 1044 taken at 11:55:20 AM CST
Picture 1045 taken at 11:56:00 AM CST
Picture 1046 taken at 11:51:08 AM CST

Again, I was only able to see the UFO when wearing my polarized sunglasses, and not having a polarized lens on my camera it appears I was unable to capture the UFO in the picture.

Although I cannot see it in the picture, it may be there, it was within the viewfinder frame, and with advanced media technology used to analyze these pictures more intensely it may be possible to reveal the UFO.

The next series of pictures are the same UFO...

it has dramatically changed its appearance...

Picture 1047 - 11:56:18 AM CST
Scott Gearen

Picture 1047 was taken approximately five minutes after the previous series where the camera was unable to capture the object and was only visible to me when wearing polarized sunglasses.

The UFO has changed into an entirely different shape and completely new appearance not seen in any of the previous pictures.

The bright area seems to be concentrated on what appears to be the top of an enclosed compartment.

Notice the area below the bright section, it appears to be a solid blue space, but by using a filter to change the color it reveals something very different.

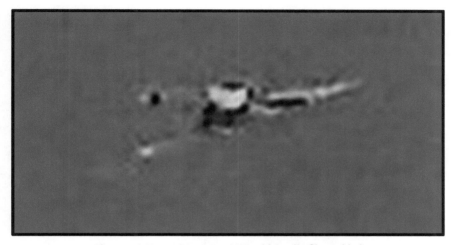

Picture 1047 - 11:56:18 AM CST -with media filter added
Scott Gearen

This is the same picture, I realized that using the media tool I could use the different filters to change the shade and tone of the colors in the picture, initially I did this to see if the UFO would be visible.

By using a different filter you can clearly see the round circle on the side of the UFO.

Without using the filter I would not have noticed the distinctive and separate sections, separated just as they would be with muntins in a modern window separating the panes of glass.

Could this round structure actually be a window, or could it be an air intake or an exhaust port...

There is no way to tell, unless someone with information about this UFO reveals it.

Regardless, this does not appear to be something that would be found in nature.

The following three pictures: 1048 through 1050 were taken over approximately 20 seconds.

Picture 1048 taken at 11:56:24 AM CST (Not Shown)
Picture 1049 taken at 11:56:36 AM CST (Not Shown)
Picture 1050 taken at 11:56:44 AM CST (Not Shown)

Again, I was only able to see the UFO when wearing my polarized sunglasses, and not having a polarized lens on my camera the lens was unable to capture the UFO in the picture.

It may be possible with advanced media technology to analyze these pictures more intensely and reveal the UFO.

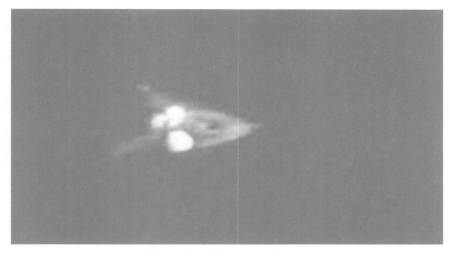

Picture 1051 - 12:31:08 PM CST
Scott Gearen

Picture 1051 was taken 34 minutes and 24 seconds after picture 1050.

Since the UFO was going though metamorphosis cycles that were lasting about one minute each there may have been up to 30 or more of these dynamic cycles the UFO experienced since the last picture.

At first glance the UFO appears to be a rocket taking off, with the three main thruster engines burning brightly as it accelerates away.

Although it appears the UFO is moving at an extremely high rate of speed, it is not leaving its position, the appearance of movement is all within the UFO itself.

The UFO does appear to be more luminescent as if it is undergoing sublimation and deposition as it changes between solid and gaseous states.

Picture 1052 - 12:31:12 PM CST
Scott Gearen

Picture 1052 was taken four seconds after picture 1051.

This is an amazing picture. This may be the "OMG" picture of the series.

I do not know what it is or what I witnessed and captured through the lens of my camera; it appears this UFO has just undergone a metamorphosis.

The UFO has turned itself into the largest and most amazing butterfly I have ever seen.

However, as I was taught at various survival schools... "Nature does not make straight lines", and the edges of these wings are very straight and appear to be manufactured by someone or something.

Is this Extraterrestrial Technology, a Weather Balloon, or... is it a "Sky-Stingray"...

Picture 1053 - 12:31:18 PM CST
Scott Gearen

Picture 1053 was taken six seconds after picture 1052.

The UFO continues to change shape. It is difficult to determine what the shape is, but the object I identified as the "eye" is prominently positioned at the six o'clock position.

The UFO is a mosaic of colors and shapes within the outer edges, the colors and bright areas are unsettled and movement is ever present.

It is apparent the metamorphosis is not over.

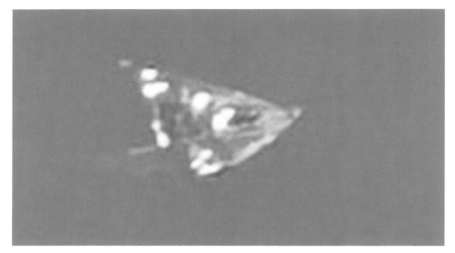

Picture 1054 - 12:31:26 PM CST
Scott Gearen

Picture 1054 was taken eight seconds after picture 1053.

The shape from the previous picture has slightly changed, and the bright areas, that resemble solar panels continue to fluctuate in both size and intensity.

The edges are still present and it appears a "wing" like structure, as the picture is viewed, is slightly upturned and the bottom side appears to disappear as it blends perfectly with the surrounding blue sky.

Pictures 1055 and 1056 were taken two minutes after picture 1054.

Picture 1055 taken at 12:33:48 PM CST (Not Shown)
Picture 1056 taken at 12:34:44 PM CST (Not Shown)

I was only able to see the UFO when wearing my polarized sunglasses, and not having a polarized lens on my camera the lens was unable to capture the UFO in the picture.

It may be possible with advanced media technology to analyze these pictures more intensely and reveal the UFO.

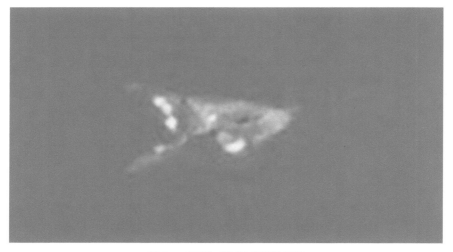

Picture 1057 - 12:34:50 PM CST
Scott Gearen

Picture 1057 was taken six seconds after picture 1056.

The UFO was previously undetectable without a polarized on my camera, and I am now able to capture an image in only six seconds from the previous picture.

The UFO seems to be adjusting itself to the light spectrum's in our atmosphere, at times I can see it with my normal eyesight and at other times only with the aid of a polarized lens.

There are still significant changes occurring within and to this UFO, the "wing" remains curled and upturned, and now a prominent dark area has formed on the topside.

There is no way to identify what this UFO is...
unless you know what it is.

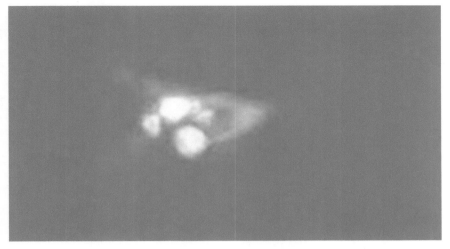

Picture 1058 - 12:35:00 PM CST
Scott Gearen

Picture 1058 was taken ten seconds after picture 1057.

The UFO has gone from a "static" appearance and once again it appears to be moving at an extremely high rate of speed, even though it remains in the same general area.

With no forward projection to keep it aloft, or overhead rotors as to enable it to hover...

How is the UFO defying gravity?

Are the colors and hazy appearance due to a lighter than air gas, such as helium?

Is this a chemical reaction?

Are the bright areas an energy source?

What is it and where did it come from?

Picture 1059 - 12:35:06 PM CST
Scott Gearen

Picture 1059 was taken six seconds after picture 1058.

The outline of the UFO is beginning to appear, and the brightest areas have changed shape, location, and intensity.

The UFO continues to display what appears to be a "wing shape" as observed from my location, with only the solid blue underside visible.

The colors within the UFO continue to fluctuate, with a distinct linear dark spot which appears sporadically on the upper area of the UFO.

This dark area changes its shape and location on the UFO just as the bright areas do.

This UFO presents more questions than answers...

Picture 1060 - 12:35:14 PM CST
Scott Gearen

Picture 1060 was taken 14 seconds after picture 1059.

It appears the grouping of smaller bright spots are positioned on the underside of the UFO. However, upon closer examination it is evident the bright areas are on the top side of the UFO.

This area appeared due to a change in the position of the upturned "wing" exposing an area not previously seen.

Why is this "wing" photographed in different positions in each consecutive picture? Is the "wing" flapping, just as a bird would to keep it aloft.

Could this UFO maintain its position with an occasional flapping motion to stay aloft?

There is no way to know what it is, but it is highly unlikely this UFO was not noticed on radar. This UFO was in a close proximity to a major USAF installation and civilian flight paths.

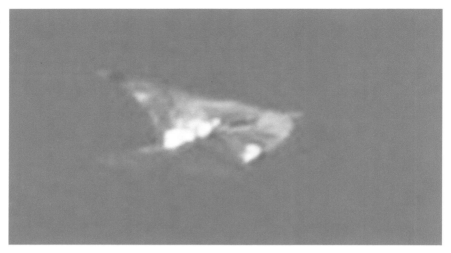

Picture 1061 - 12:35:30 PM CST
Scott Gearen

Picture 1061 was taken 16 seconds after picture 1060.

The bright areas noted in the previous picture are almost hidden by the upturned "wing" which is slightly higher in this picture partially blocking the view.

The colors of the UFO seem to be more "fluid", more prominent, and vary in intensity.

My guess is the colors are some type of gas such as helium, neon, argon, and krypton, etc. which are used in decorative lighting, known as "neon" lights.

Is this the same type of "lighting" created by "Rainbow Jellyfishes", Could this be an Underwater UFO?

Could this be what has been referred to as "Swamp-Gas?"

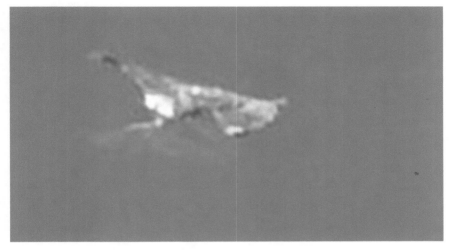

Picture 1062 - 12:43:36 PM CST
Scott Gearen

Picture 1062 was taken 14 minutes and 6 seconds after picture 1061.

The UFO appears to be in a relaxed phase and continues to maintain this same general appearance for several minutes.

There are several areas that have remained mostly unchanged throughout the series of pictures. The round dark spot on the lower right of the UFO, and what appears to be one protruding on the opposite side. Is it possible this dark area and the one on the opposite side are "eyes" similar to those found on a stingray?

Could this UFO be similar to a whale shark? A large living creature that has come down from extreme high altitudes to feed? This area where it is "hovering" is only a few miles inland from the Gulf Coast, air currents could be lifting Plankton type organisms from the Gulf into the jet stream, and this "Sky-Ray" is hovering so the air currents bring its meals to it.

Picture 1063 taken at 12:44:08 PM CST (Not Shown)

Pictures 1063 was taken 34 seconds after picture 1062.

I was only able to see the UFO when wearing my polarized sunglasses, and not having a polarized lens on my camera the lens was unable to capture the UFO in the picture.

It may be possible with advanced media technology to analyze these pictures more intensely and reveal the UFO in the picture.

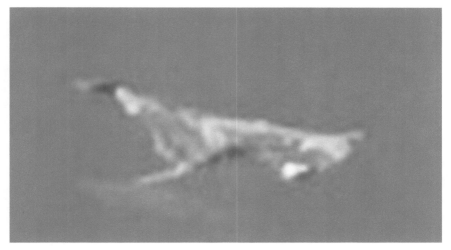

Picture 1064 - 12:44:16 PM CST
Scott Gearen

Picture 1064 was taken eight seconds after picture 1063.

The UFO continues to remain almost stationary, the bright areas have almost disappeared.

The UFO has darker colored areas that shift locations and sizes and seem to be fluctuating with various colors and shades.

The movement of these changing colors and markings the UFO is exhibiting are the same characteristics commonly found in stingrays, and many more sea creatures to camouflage themselves.

This UFO seems to have more in common with the "alien" creatures that are still being discovered in the depths of the oceans.

Based on what I am observing, and my analysis of these pictures, I am convinced this UFO is capable of existing in any environment: space, land, or sea.

Picture 1065 taken at 12:44:28 PM CST (Not Shown)

Pictures 1065 was taken 12 seconds after picture 1064.

I was only able to see the UFO when wearing my polarized sunglasses, and not having a polarized lens on my camera the lens was unable to capture the UFO in the picture.

It may be possible with advanced media technology to analyze these pictures more intensely and reveal the UFO.

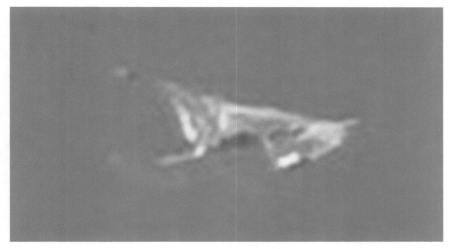

Picture 1066 - 12:44:36 PM CST
Scott Gearen

Picture 1066 was taken eight seconds after picture 1065.

The UFO seems to be maintaining the same shape over the last several minutes, the smaller bright area on the lower right side with the black center, as I have identified as an "eye" remains constant.

The changes appear to be coming from within the UFO, the colors are becoming more vibrant as they move around the UFO... possibly heating and cooling for a new shape.

The following three pictures were taken over approximately 20 minutes.

Picture 1067 taken at 12:44:46 PM CST (Not Shown)
Picture 1068 taken at 1:11:16 PM CST (Not Shown)
Picture 1069 taken at 1:11:32 PM CST (Not Shown)

Picture 1067 was taken ten seconds after picture 1066, followed by pictures 1068 and 1069.

The UFO now appeared to be further away and more difficult to see. I was only able to see the UFO when wearing my polarized sunglasses.

I continually doubled checked each time I took a picture the lens was directed at the UFO, but without a polarized lens the camera was unable to capture the image.

While analyzing the pictures I experimented with the settings and different colored backgrounds to determine if the UFO would appear in the pictures where I could not see it.

I determined the UFO was easier to identify in some pictures, such as 1047, but still was not able to be seen in all pictures.

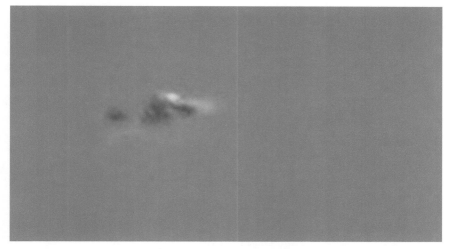

Picture 1070 - 1:12:36 PM CST
Scott Gearen

Picture 1070 was taken one minute after picture 1069.

Many of the pictures I took of the UFO were only visible through the polarized lens of my sunglasses and were not being recognized by the lens of my camera... amazingly this time it was.

Fortunately picture 1070 was captured on my camera, and amazingly it has returned to the shape captured in picture 1047.

The UFO appears to be moving further away, until I cannot see it with or without the polarized sunglasses.

The shape of this UFO now looks to me like a cross between the Virgin Galactic SpaceShipTwo and the Star Trek Enterprise.

The next 12 pictures were taken over a span of approximately 5 minutes.

Pictures 1071 through 1082 (not shown), were taken from 1:12:50 to 1:17:38 PM CST

I could see the UFO only through my polarized sunglasses, I continued to double check each time before pressing the shutter to ensure the camera lens was focused where the UFO was.

The camera was focused, the UFO was there, I could see it when wearing my polarized sunglasses, it was not visible with regular eyesight and the lens on my camera was unable to detect the UFO without the aid of polarization.

Same camera, same UFO, I did everything the same way each time I took a picture, the pictures were taken consecutively, some only seconds apart... but the image was not captured by the camera each time... or, maybe it was...

Why didn't the camera "capture" the image every time?
What prevented my eyes and the camera from seeing the UFO?

The ultimate camouflage?

Picture 1083 taken at 1:17:50 PM CST
Scott Gearen

Picture 1083 is the last picture of the UFO captured by my camera it was taken 12 seconds after picture 1082.

The "point" or "nose" area of the UFO appears to be shaped like a military jet aircraft, there is a bright green color surrounding the outer edges of the "nose", inside the green colored area, in the interior of the UFO is a strange dark, rectangular area, with a distinctive red color in the center.

I am well past OMG, and have gone straight to WTF!!!

Another distinctive area of the UFO is the dark blue "wing like structure" that appears to stick out of the main body of the UFO at a 90 degree angle, it has a long straight edge, that looks like it would be part of a fixed wing aircraft.

There is also another dark blue color that can be seen in the upper left section of the UFO.

I took four more pictures following picture 1083, and they were not captured by my camera.

I was still able to see it, but only while wearing my polarized sunglasses. Maybe I wasn't supposed to see it...

Starting with picture 1063 and ending with picture 1088, a total of 26 pictures were taken of the UFO, I was able to see it each time I took a picture and only four of the pictures included the image of the UFO.

The UFO was managing to hide itself from detection by my eyes and my camera. Maybe it was always able to "hide" itself, and only through some type of malfunction of its system was I able to see it.

Maybe it is "hiding" itself now... maybe it is up there now and it is undetectable to our technology.

Whatever the UFO was doing it was able to prevent me from seeing it continuously with my normal eyesight and my camera appears to have been unable to capture it.

It may be possible with advanced media technology to analyze the photos that only show a blue sky and expose the UFO in the picture.

| 4 |

All Options Are On The Table

The truth is always an option... regardless of how strange it may be

This list of options is not all inclusive, they may seem far fetched, and even laughable, but no one has positively identified this...

Unidentified Flying Object

"All options are on the table"

Some possible explanations of the UFO are:

- A highly secret project of the United States government utilizing a recovered Extraterrestrial Space craft which has been re-engineered.

-

- A highly secret project of the United States government working in conjunction with Extraterrestrial Beings, who are providing materials and supervision for this project.

-

- An alien spacecraft from another galaxy operating on its own, and possibly observing humanity.

-

- An unknown "cryptologic" creature that primarily lives in oceans around the world and is capable moving in our out of the ocean, to include in space, on land or in the sea.

-

- An unknown "cryptologic" creature that primarily lives in deep space, outside of our atmosphere, and is capable moving in our out of earthly environments at will.

-

- "Solar Warden" – a Top-Secret, anti-gravity space vehicle from this "rumored" Government program. Rotating personnel for their "Tour of Duty".

-

- A Magnetohydrodynamic drive vehicle... such as "Wingless EFE Touring Craft MHD".

-

- The "Black Knight", or another artificial satellite circling the earth. Such as the unidentified object photographed from STS88.

-

- A "Rigid Hull Airship".

-

- "Victor One" supposedly the Spaceship of "Valiant Thor". (Maybe...)

-

- Inter-dimensional Travel - An Unidentified Object traveling between dimensions.

-

- Time Travel - Our future returning for an update on the human race.

-

- No doubt, there are other possibilities, but just maybe one of the options listed here is the UFO. Regardless, whatever it is, these pictures captured by my camera represent an advanced technology that is light years ahead of anything 99.9 percent of humans are currently aware of or using.

| 5 |

Fact vs Fiction

It's All Fact

FACT: I saw an Unidentified Flying Object.

FACT: I have photographic evidence to prove it.

FACT: I have never seen anything like it before... ever.

FICTION: There is none... *it is all Fact...* 100 percent guaranteed.

Someone, somewhere will see these pictures and know what this is.

But until they provide information as to what this is, it is no doubt, the most amazing **Unidentified Flying Object** you have ever seen.

I challenge anyone to prove these pictures are not authentic, and I welcome anyone with knowledge of this UFO to contact me... I want to go for a ride... again ;)

ET had the technology to go home anytime it wanted to.

ACKNOWLEDGEMENTS

Teachers. In life, in the classroom, and in space... you are essential.
If you can read this, thank one.

To the UFO's...

We know you are out there...

Scott Gearen is a retired USAF Pararescue Specialist (PJ) with over 2000 hours of flight time in USAF helicopters, he has an in-depth knowledge and awareness of what he sees in the sky. On 30 April 2019 it became clear very quickly the object he saw in the sky was not a typical aircraft. Fortunately, he had a high-quality camera and captured the most amazing and only pictures like these ever. As an amateur photographer, Scott has captured, the first, and most detailed pictures of an Unidentified Flying Object in history.

Scott Gearen grew up in rural Florida where he often watched parachutists jumping out of perfectly good airplanes at Z-Hills, and wished it was him. The urge to jump out of airplanes didn't go away and ultimately he found his way to a recruiting office where he enlisted in the United States Air Force with the single minded intent to go straight into the Pararescue career-field where he would get to jump out of airplanes, which he did many times over his 22 year career as a PJ.

You could not have a better witness of an event of this magnitude, to report with extraordinary accuracy everything he witnessed, and to back it up with detailed photographic evidence. Scott was in the right place at the right time to see this UFO and provide such a detailed report.

CPSIA information can be obtained
at www.ICGtesting.com
Printed in the USA
BVHW021036310820
587679BV00001B/1